Subliminal Success

How to Harness the Power of Your Subconscious Mind

Wil Dieck

TMT Publishing

Copyright © 2020 by Wil Dieck

All rights reserved. No portion of this book may be reproduced mechanically, electronically, or by any other means, including photocopying, without written permission of the publisher.

It is illegal to copy this book, post it to a website, or distribute it by any other means without permission from the publisher.

Limits of Liability and Disclaimer of Warranty

The author and publisher shall not be liable for your misuse of this material. This book is strictly for informational and educational purposes. Warning – Disclaimer: The purpose of this book is to educate and entertain. The author and/or publisher do not guarantee that anyone following these techniques, suggestions, tips, ideas, or strategies will become successful. The author and/or publisher shall have neither liability nor responsibility to anyone with respect to any loss or damage caused, or alleged to be caused, directly or indirectly, by the information contained in this book.

Copyright © 2020 Wil Dieck

ISBN: 978-1-956169-14-0 (eBook)

ISBN: 978-1-942844-27-3 (Paperback)

ISBN: 978-1-956169-15-7 (Hard Cover)

Wil Dieck

wil@mindfulmindhacking.com

San Diego, CA

Acknowledgments

EVERY JOURNEY STARTS WITH a single step, and this book wouldn't exist without all the steps taken by those I owe so much thanks to.

First and foremost is my lovely partner Lynette Seid who has been there for me like no other. Your love lifts me up every day!

Next are my two children Vanessa and Samuel Dieck from whom I have learned so much about what it means to be a better man and to be resilient to the challenges that comes my way.

My parents Arne & Lealia come next on my thank-you list. While long gone, their influence still lives within me today, propelling me towards even greater accomplishments.

Lastly, you, my magnificent reader.

May the words you read here embed themselves so deep into your subconscious that they drive you beyond mere reading and toward achieving what you want out of life. Maybe together we can change what we do into something worth mentioning for future generations to come.

Again, thanks for choosing to read this book!

All the best!

Wil Dieck

Black Belt Breakthroughs, LLC

Wil@mindfulmindhacking.com

https://mindfulmindhacking.com/

Contents

PART I How the Subconscious Works	1
1. Your Hidden Power	5
2. What Is the Subconscious Mind?	11
3. The Power of Your Beliefs	17
A Sincere Request	25
4. Understanding Self-Criticism	27
5. Other Negative Influences	37
Part II Take Control of Your Subconscious Mind	41
6. The Power of Affirmations	44
7. Visualization	53
8. Meditation	72
9. Hypnosis	83
10. Self- Hypnosis	93
11. Neuro Linguistic Programming	100

12. Mindfulness	112
Discover Your Path to Personal Transformation - Free Course!	127
13. How to Know If Your Reprogramming Efforts Are Working	129
14. Consistent, Persistent Reinforcement	132
About the Author	136
Other Books by Wil Dieck	138

PART I How the Subconscious Works

"Your subconscious mind does not argue with you. It accepts what your conscious mind decrees. If you say, 'I can't afford it,' your subconscious mind works to make it true. Select a better thought. Decree, "'I'll buy it. I accept it in my mind." - **Dr. Joseph Murphy**

Do you drive? What I mean is, can you drive a car?

If you can, think about the simple task of changing lanes while driving a car.

Right now, close your eyes, grip an imaginary steering wheel, and go through the motions of a lane change. Imagine that you are driving in the left lane, and you would like to move over into the right lane.

Before reading on, actually try it.

That is a fairly straightforward task, right?

When you did the maneuver, you probably held the steering wheel straight, then cranked the wheel slightly over to the right for a moment and followed that by straightening it out again.

Easy–right?

Well, not quite.

If you did what I explained above, you, like almost everyone else in the driving universe, got it exactly wrong.

The motion of turning the wheel to the right for a tiny bit, then straightening it out again, would steer you completely off the road. You just turned your car from the left lane into a ditch.

The correct motion for changing lanes is turning your steering wheel slightly to the right, then back through the center, and finally turning the wheel an equal distance to the left side. After that, you straighten out the wheel.

The next time you drive, you can verify the maneuver for yourself. It's such a basic, simple task that you have no problem getting it done every time you get into the car. But when you are forced to access the process consciously, you drive into the ditch.

This is an illustration of the power of your subconscious mind.

It does incredible tasks that you are not even consciously aware of doing!

You have no conscious awareness of the vast majority of your brain's ongoing, everyday activities. Not only are you not aware of them, many of them are tasks you would not want to interfere with.

From riding a bike to tying your shoes and guiding your car home while listening to a game on the radio, these tasks are all done subconsciously.

But when your subconscious starts to steer other areas of your life, it also has a dark side. We'll look at this more a little later.

First, let's dive deeper into your subconscious mind.

Chapter 1

Your Hidden Power

"People are anxious to improve their circumstances but unwilling to improve themselves, they therefore remain bound." - **James Allen**

Your subconscious mind has immense power. In fact, top inventors, athletes, and leaders from around the world already know the power of the subconscious mind. Many, if not all, use the methods you'll find in this book to achieve their goals and dreams.

They understand, just as you soon will, that just like steering your car, your subconscious mind steers what you do on a daily basis. This, in turn, controls how you experience your life's events and your future.

Your subconscious mind is the primary control center of your whole body. It's like you're one of those control centers of a big production plant you see in the movies. Your brain takes in information and then sends out instructions to the rest of your body.

Whatever your subconscious sets out to, affects your entire system– inside and out. Some parts, like heartbeat and breathing, operate all by themselves and are thought of as other than conscious processes. Many have learned to control even these processes.

For example:

As long ago as 500 C.E., Tibetan monks would sit naked on the edge of a 12,000-foot mountain before an icy stream all night. This was not to punish them, but rather as a test. The monks were there to demonstrate what they had learned.

If the freezing cold wasn't enough, they wrapped the monk's bodies continuously in bed sheets soaked in ice water.

For the "normal" person, this deadly mix of freezing temperatures, relentless winds, and painful ice would cause hypothermia, frostbite, and almost certain death.

Yet, because of their training, the monks utilized the power of their subconscious minds to raise their core body temperature. As a result, each survived with zero pain or illness. All of this was thanks to the power of their subconscious minds.

Even today, monks go through this same initiation ceremony with identical results. They, like all the monks before them, have trained their subconscious minds to create warmth inside their bodies, despite what's going on in the world outside them.

Modern Day Uses of the Subconscious Mind

"Whatever we plant in our subconscious mind and nourish with repetition and emotion will one day become a reality." — **Earl Nightingale**

What you believe you can or cannot accomplish is also a deep part of your subconscious thought process. If you talk to nearly any Olympic or professional athlete, they will all tell you that they "practice" their sport in their minds. They tap into the subconscious because they understand that's where their "instinctual" or habitual moves come from. This way, when they need to use them, the moves come "naturally". Using mental practice is how they program their subconscious minds for success.

The great news is you can learn these skills too! That's what this book is about; introducing you to the skills you can use to harness the power of your subconscious mind.

What you need to understand is that when you know how to operate these processes so you can steer your subconscious on command, the possibilities it can give you are endless.

On the other hand, if you don't know how to operate these processes, they can hold you back from realizing these possibilities.

Everything in your life, from the types and quantity of the food you eat, the level of income you earn, the habits that control your actions and even how you react to stressful situations, directly results from these subconscious processes.

Your life and everything you accomplish or don't is a direct result of your subconscious thoughts and beliefs. This is why many experts argue that this is the most powerful force in your life.

Your Own GPS System

"It is psychological law that whatever we desire to accomplish we must impress upon the subjective or subconscious mind." - **Orison Swett Marden**

Your subconscious mind acts as a map similar to the map in your car's GPS system. Your car's GPS system works well as long as the map it has is up to date. However, if your car's GPS system's map is out of date, it will contain routes that lead you along the slow path or can even guide you to the wrong destination.

This will result in you being constantly lost or going down paths that take a long time to get you where you want to go. If what you're looking for has moved, you may not get there at all. For your car's GPS system to work properly, it needs constant updating.

Your subconscious mind works similarly to your car's GPS system. To work, your maps need to be up to date.

If you have up-to-date maps, it will take you in the direction you want to go. But if the map you're using is not up to date, if the information your subconscious mind is relying on is not accurate, then it will take you in directions you may not need or want to go.

For your subconscious mind to work properly, it needs to be programmed with the most updated routes. Routes that will get you quickly to the destinations you desire.

Chapter 2

What Is the Subconscious Mind?

"Our subconscious minds have no sense of humor, play no jokes, and cannot tell the difference between reality and an imagined thought or image. What we continually think about eventually will manifest in our lives." - **Robert Collier**

The "subconscious" is the part of your mind that operates below your normal level of waking consciousness. Again, psychologists often refer to as being "other than conscious" processes.

If you could look inside at your mind, it would appear to be like an iceberg. The conscious mind is like the part of the iceberg, about 10 to 12 percent, that you can see above the surface.

But the bulk of the mind, about 88 to 90 percent, is the subconscious mind. This part is below the surface, unseen, and usually unnoticed.

For example, when you breathe, you don't consciously direct every muscle that expands or contracts to pull in and push out the air from your lungs. When you walk, you don't consciously tell your muscles to pick up each foot and put it back down as you step along your path.

Your subconscious, or other than conscious processes, handles all this work for you.

In the same way, your subconscious directs your behaviors and habits. These other than conscious processes even cause conscious thoughts and physical sensations based on that same data.

Where Does This Programming Come from?

"We cannot always control our thoughts, but we can control our words, and repetition impresses the subconscious, and we are then master of the situation." - **Jane Fonda**

Some experts will tell you that everything you've ever seen or heard is sitting somewhere in your subconscious mind. These memories, no matter how long ago, can, and often do, affect your current thoughts, decisions, and actions today.

Your existing perceptions, how you see the world, have been forming your entire life. This began when you were just an infant. As you moved through and experienced life, your subconscious mind soaked in information like a sponge.

Your subconscious mind learned many things. For example, when you cried, a magical creature that you later learned was your mother would come to feed you, change you, or burp you. It also learned that when you smiled, people would pay attention to you, especially that magical creature.

As you grew older, it began making up rules about your life. It also made rules about the people and things around you.

For example, if learning was important to your mother, she would probably give you a reward. The reward might be her saying, "you're so smart" when you learned something new. As a result, you made up the rule that "learning is good."

But if your dad punished you with words by saying "can't you see I'm busy" a couple of times as you tried to ask him a question, you made up the rule "talking to dad is bad."

While neither of these rules was absolute, your subconscious made a quick connection that you relied on to make future decisions about what actions you would take. It did this without question.

You see, while you were young, your subconscious mind rejected nothing while allowing everything to come in. This was because you didn't have any pre-existing beliefs to contradict what it perceived.

It simply accepted that all the information you received during your entire early childhood was true.

By the way, this is why there are age recommendations on video games and movies because young kids can't tell the difference between them and reality (we adults can also have those problems but that's a discussion for a later time).

The Problem with Your Early Programming

"You affect your subconscious mind by verbal repetition." - **W. Clement Stone**

This early programming resulted in you carrying around some real baggage.

Every time someone called you stupid, worthless, slow, lazy, or worse, your subconscious mind simply stored the information away for reference. It was there for you to recall anytime you wanted.

You, like so many of us, may have also received messages about your potential or limitations based on your physical abilities, skin color, gender, or economic status. These too are in your subconscious for quick reference.

Here's the real problem. By the time you were seven or 8 years old, your subconscious mind had been programmed with a great number of beliefs, both false and true, based on all that input from people in your life, television shows and movies you watched, along with many other environmental influences.

How Does This "Old" Programming Affect You Now?

"Only one thing registers on the subconscious mind: repetitive application - practice. What you practice is what you manifest."
- Fay Weldon

Your subconscious mind contains vast stores of information. These are the stores of information you've been acquiring your entire life. They are also much more than your conscious mind could ever handle.

As an adult, you may think that you can simply forget or discard any hurtful or untrue messages you absorbed during your early life. If only it were that easy. It's not easy because one of the most fundamental ways you give your experience meaning is through your beliefs.

Your beliefs have been stored in your subconscious mind. They act to reinforce your motivation to work hard and accomplish the goals you set for yourself. They also act to sabotage you if it has been trained to believe you are unable.

These beliefs are stored deep in your subconscious. They affect how you habitually think. They can also automatically block any attempt you make to make a change in your life.

The only time you actually become aware of the effect of your negative subliminal programming is when it limits your progress in creating a balanced, successful, and productive life.

This is why it's so important to learn the skills in this book.

Chapter 3

The Power of Your Beliefs

"If you think you can you can, if you think you can't you can't." - **Henry Ford**

Let me ask, have you ever wanted something so bad that you willed it to happen?

For example, you wanted a promotion and it suddenly appeared, or you wanted a new car, or a new boyfriend or girlfriend and it happened?

Of course, you have! This has happened to us all in one form or another many times in our lives.

Here's what you need to understand. These things don't happen by chance. They are not miracles, either. By sheer will, through the power of your subconscious mind, you made these things happen.

As we've discussed, the mind is the main control center of your whole body. Whatever it sets out to do affects your whole system – inside and out. When you have the skills, you need to operate and steer it to your command. Then the possibilities are endless.

One of the best examples comes from running a mile. At one time, people thought it was impossible for a human to run a mile in less than four minutes. This was until May 6, 1954, when Roger Bannister ran it in 3 minutes 59.4 seconds. After Sir Roger did it, hundreds of people of all ages followed suit. In fact, Eamonn Coghlan was 40 years old when he broke the four-minute mile barrier.

What does this have to do with you?

Have you ever tried to achieve a goal and you found yourself sabotaging yourself at every turn, find yourself saying "I knew that would happen" when things went wrong?

This happens because of your limiting beliefs. This same thing was happening to the runners before Roger Bannister broke the 4-minute mile. When they tried, they knew they would fail.

The good news is you, like those runners, can change your beliefs. You are not defective. You are not doomed to fail. You just have OMS (old map syndrome).

OMS is when your old programming (maps or beliefs) conflict with the new conditions you want to create.

You need to rewire your subconscious thinking. You need to reprogram your beliefs. Your new programming will change what you believe you can do.

Reprogramming is a skill anyone can learn. This means you can learn to rewire your thinking and change your beliefs!

You can adjust them to help you achieve just about anything you want IF you first take the time to learn how to reprogram your subconscious mind!

Your Programming Is Ongoing

"If you accept a limiting belief, then it will become a truth for you." - **Louise L. Hay**

Before we talk about how to reprogram your subconscious mind, it's important to know that the programming we've talked about is still going on today. Every experience you have causes you to come to certain conclusions about those events. These conclusions will be stored in your subconscious and will guide your future actions.

For example, maybe you were rejected by someone you cared about like a boyfriend or girlfriend, or even a husband or wife. What kind of messages would your subconscious mind look for?

Well, it really depends on what you ask it to go looking for.

If you ask, "Why does this always happen to me?" – your subconscious mind (that clever detective) would immediately go hunting through your memories and find other examples of rejection, like that time your best friend dumped you to hang out with the more popular kids.

From here, it might draw a conclusion that you're somehow unworthy or unlovable and deserve to be rejected. Your subconscious mind has found the answer you were looking for.

But let's say today you have an experience that conflicts with an already established belief. Instead of taking on a new belief, your subconscious mind will do everything in its power to either reject it or reframe it, so it fits in with your existing view of reality.

Let's examine a different example.

Many people hold the belief that they are unattractive. Maybe you are one of them. Now let's say a person you feel is exceptionally attractive expresses an interest in getting to know you better.

Some people might think that this must be a joke or a cruel trick. If you do, since you already believe that you're unattractive, you probably won't be able to believe that this very attractive person could find you attractive.

Beneath your conscious awareness, your antenna goes up and your subconscious mind shouts, "No way! This person is way too attractive to be interested in me, something isn't right here..."

As a result, you'll either outright reject the idea of having a relationship with that person or, if you do begin a relationship, you'll end up subconsciously sabotaging it.

Either way, you'll actively sink any chance of developing what could have been a great relationship. This is the same thing that happens when you struggle to achieve any goal.

When you set a goal that you think is outside of your abilities, you sabotage it. As a result, eventually you begin to believe that you aren't capable or worthy of success. You come to expect failure and end up failing over and over again!

You can probably imagine many other situations where your subconscious mind limits you. Much of the time, it's a direct result of your inner critic.

Your Inner Critic

"Self-criticism and negative thoughts about yourself will attract people who reflect this back to you, showing critical behavior and can abuse you physically." - **Hina Hashmi**

Your inner critic is lurking inside your subconscious mind. Each of us has an inner critic. I'm sure you are acquainted with your own.

This critic is exceptionally good at reminding you of your faults and failings. The biggest problem is, left unchecked, your faults and failings are all your inner critic will let you focus on.

After a while, those self-critical thoughts can become overwhelming. They consume your thinking. Night and day, they will be there. After a while, it can be difficult to focus on anything else.

For most people, negative self-talk is an ongoing dialogue running around inside your head. This dialogue focuses on your flaws and self-perceived weaknesses.

The worst part?

Most of the time, these thoughts have no factual basis. But, since you're the one saying these things, you tend to believe them.

Here's something you need to remember. Self-criticism isn't always bad. In fact, there are times where self-criticism can be downright useful.

For example, when examining a past mistake to find a solution so you won't do it again. This is an excellent use of self-criticism.

The problem is, most of the time, your self-criticism is simply creating unnecessary pain and suffering. The thoughts your self-critic produces increases your risk of depression, anxiety, stress, and other mental health issues. These critical thoughts also negatively affect your self-esteem and self-confidence and, as a result, your relationship with yourself and others.

Fortunately, you can learn to overcome these negative thoughts. You're going to learn an excellent process of how to do just that in this book.

But first, let's look at where your self-critic comes from and how it affects you.

A Sincere Request

Dear Valued Reader,

I wish I could address this to you personally, but our technology hasn't advanced that far yet.

With that in mind, I wanted to thank you for taking the time to read my book. I hope it brought a bit of joy or inspiration into your life. It really means so much to me that you took this journey with me.

Now I'm going to ask you for a favor.

If reading my book felt valuable to you, please consider leaving a review on Amazon (here's the link https://www.amazon.com/dp/B075JPC6TL).

or you can scan the QR code below

Your opinion is invaluable in helping me create something impactful and meaningful with my writing, not only now but continuing forward as well.

And I'm open to whatever mode of feedback works best for you - ratings, reviews comments either online or offline.

They are all deeply appreciated.

And if at any time you have questions while reading through my book, feel free drop by website https://mindfulmindhacking.com/ where can chat directly together about anything related to Mindful Mind Hacking or any of my books.

Hope to hear from you soon!

Sincerely,

Wil Dieck

Chapter 4

Understanding Self-Criticism

"Much protective self-criticism stems from growing up around people who wouldn't or couldn't love you, and it's likely they still can't or won't. In general, however, the more you let go of the tedious delusion of your own unattractiveness, the easier it will be for others to connect with you..." - **Martha Beck**

As we discussed, we all have a critical self inside of us. This critic is like another person who is pointing out our self-perceived flaws.

As pointed out, sometimes self-criticism can be a healthy way to increase your self-awareness. This type of self-criticism can lead to a better sense of self-worth.

This is because it's only when you're willing to honestly examine yourself that you can overcome areas of weakness or unwanted habits. This type of self-criticism leads to greater awareness and personal growth.

Your inner critic can also become a barrier to your success and happiness. If you are overly critical of yourself, your self-criticism can prevent you from taking risks.

Self-criticism can reduce your self-belief. It can also hold you back from expressing your opinions or your feelings about yourself or others.

Note: If you are experiencing an elevated level of self-criticism, you may wish to address the reasons behind those tendencies with a therapist or other mental health professional. A good therapist can help move you quickly to reduce your self-critical thoughts and to a higher level of self-worth.

Types of Self-Criticism

Generally speaking, you can divide self-criticism into two categories, comparative and internalized criticism.

Comparative criticism is comparing yourself to others. Internalized criticism is feeling like you can't live up to your own standards.

Because both comparative and internalized self-criticism can be harmful, let's take a quick look at each of them.

Comparative Self-Criticism

Comparative criticism is when you are constantly comparing your life to others and then finding yourself lacking. A self-critical person often forms their self-esteem on their perceptions of the way others feel about them. They also believe that others are constantly judging them.

Comparing yourself with others makes it easy to view others as superior or better than you. This can lead you down the path of thinking that you don't measure up.

Continually comparing yourself with others this way will damage your self-image. It can also ruin your relationships with your friends, family members and coworkers. Doing this will absolutely hijack your happiness and success.

Internalized Self-Criticism

Internalized self-criticism stems from a desire to be perfect. But, since nobody's perfect, this sets you up for failure. You will not always measure up to a perfect standard in one way or another.

This can result in you feeling you are not successful, even when you are. One common example I have seen first-hand comes from years of teaching college students.

A student is used to receiving A's on all their assignments and receives an A minus on a test. After class, they approach me and ask what they've done wrong. They seem stressed and act as if they have failed.

This student has a belief that anything less than perfection is failure. The problem with this obsession with being perfect is it can quickly become overwhelming. It can mentally paralyze you with the fear of not reaching your standard of perfection.

Now that we've examined the two distinct types of self-critical thoughts, let's look at how you can begin to effectively address their issues.

Where Do Self-Critical Thoughts Come From?

"It is not the critic who counts; not the man who points out how the strong man stumbles, or where the doer of deeds could have done them better. The credit belongs to the man who is actually in the arena ... who strives valiantly, who errs and comes short again and again, because there is not effort without error..." - **Theodore Roosevelt**

In many cases, the root cause of self-critical thinking and their associated beliefs are a result of past experiences. Parents who are controlling and rigid can negatively affect a child's self-worth and lead to a child's negative self-perception.

Children who feel rejected because of a parent's lack of warmth and compassion, or who are constantly criticized, are also negatively affected. Because their frame of reference is so critical and negative, these children are much more likely to grow up deeply critical of themselves and others.

This type of self-critical thinking can lead to the three most common negative thoughts.

The Three Most Common Negative Thoughts

"Those who improve with age embrace the power of personal growth and personal achievement and begin to replace youth with wisdom, innocence with understanding, and lack of purpose with self-actualization." - **Bo Bennett**

While there are many more to choose from, the following are the three most common negative thoughts.

"I'm Not Good Enough"

"I'm Not Smart Enough."

"I Don't Belong Here"

To change these types of thoughts you need to reframe them. In this section, you're going to learn how to reframe these self-critical thoughts before they make your life even more difficult.

Reframing Your Inner-Critic

"For any single thing of importance, there are multiple reasons." - **M. Scott Peck**

How can you learn to reframe (We'll take a deeper dive into reframing more in the NLP section) the negative thoughts your inner critic is telling you?

You can do this using the Five-R Process.

The 5-R Process

The book, Mindful Mastery, introduces the reader to the 5-R process. You can use the 5-Rs to capture and reframe your self-critical thoughts.

The 5-R process consists of five steps. The steps are:

1) recognize,

2) record,

3) remind,

4) replace and

5) repeat.

Let's look at each of these steps.

Recognize

The first thing you have to do with any negative habitual thought is to become aware of it.

Often, thoughts like "I'm not good enough", come up so often that we don't even pay conscious attention to them. But just because you are not hearing it consciously, this doesn't mean it is not affecting you.

One of the things I teach my clients is how to become more aware of these types of thoughts. I do this by teaching them simple mindfulness techniques. You can learn more about how you can begin to practice mindfulness in chapter 12.

Record

The next step is to record what your self-critic is telling you. You do this in the moment as soon as these thoughts come up. You can use your smart phone or just a simple notepad to do this.

In the evening, you're going to transfer your self-critical thoughts into a journal and analyze them.

Remind

After you have recorded and analyzed your self-critical thoughts, you'll need to come up with reminders about why they aren't really true. For example, just because you haven't stayed on that exercise program for more than a week doesn't mean "I'm not good enough."

During this step, recall times when you were good enough. For example, when you graduated from college or when you completed a complicated project at work.

Now you're going to turn these memories of past successes into replacement phrases.

Replace

The replacement step is simply replacing your old pessimistic, negative critical self-talk with positive, optimistic self-talk. Using the example from above, you might say, "I'm good enough to handle anything that comes my way. Especially my health plan."

For more information about how to create optimistic self-talk, look at chapter 6, "Affirmations."

Repeat

The repeat step is turning your new replacement phrase into your habitual self-talk. You do this by creating a new habit. This simply means doing the new behavior enough times that it happens automatically. You can learn more about how to create new habits in my book "Secrets of the Black Belt Mindset: Turing Ordinary Habits into Extraordinary Success".

Using this simple five-step process, you can learn to reframe your self-critical thoughts and turn them into a powerful tool for your success!

In Part II we'll examine other methods you can use to silence your subconscious inner critic.

Right now, let's look at other negative subconscious influences.

Chapter 5

Other Negative Influences

"You can accomplish anything you wish that is not contradictory to the Laws of God or man, providing you are willing to pay a price." - **W. Clement Stone and Napoleon Hill**

In a little while, we'll begin looking at some different ways to overwrite the limiting or damaging messages that your inner critic has been storing in your subconscious mind.

First, let's look at some other negative influences that affect you.

Environmental Influences

"You must be the change you wish to see in the world." - **Gandhi**

As you've seen, your subconscious mind is constantly absorbing information from your environment. As a result, it draws conclusions and forms beliefs based on that information.

Here's how it works. If your daily environment is filled with drama and negativity, those types of messages are being absorbed into your mind.

To begin to change this, ask yourself, "How is my environment affecting my subconscious mind?

All That News Isn't Good for You

"The world is a great mirror. It reflects back to you what you are. If you are loving, if you are friendly, if you are helpful, the world will prove loving and friendly and helpful to you. The world is what you are." - **Thomas Dreier**

Some people watch hours of news and read newspapers and magazines all the time. As they do, they become agitated, and their blood pressure rises.

If this is you, stop doing that!

One of the best things you can do is strictly limit the negativity you allow yourself to be exposed to in the media from this moment on.

While you do want to be informed, avoid watching the news too often unless it's absolutely necessary. Read only one or two magazines or newspapers that cover the news and only read them weekly.

You Are Who You Associate With

"You are the average of the five people you spend the most time with." - **Jim Rohn**

Another thing to do is avoid spending too much time with "toxic" people. You know the kind I'm talking about. These people always have something negative to say about the world or you.

Commonly known as "energy sappers", they sap the positive energy from everyone and everything around them.

Instead of spending your time with negative news and people, seek out positive information to read and watch. Spend most of your time with positive people who are focusing on doing something that builds up the world with their lives.

Once you get into this habit, what you'll find is that, over time, by exposing your mind to increasingly uplifting and encouraging messages you will change the way you see yourself and your future potential.

Now that you have a better understanding of how other influences affect your subconscious mind, let's look at more methods you can use to reprogram it for success.

Part II Take Control of Your Subconscious Mind

―――*ele*―――

Experts, including mystics and scientists, agree that your main control center, your subconscious mind, can be controlled and harnessed through affirmations, visualization, mindfulness, meditation, visual suggestion, and mind exercises such as hypnosis and NLP.

Here's a quick preview of these processes.

Affirmations are positive statements that you repeat to yourself, consciously directing your thoughts and beliefs towards desired outcomes.

Visualization is a powerful technique that involves creating vivid mental images of your desired outcomes.

Mindfulness and meditation practices cultivate present-moment awareness and focus. By training your mind to be fully present and non-judgmental, you can develop a heightened sense of self-awareness and gain greater control over your thoughts and emotions.

Hypnosis and NLP are techniques that aim to access and reprogram the subconscious mind on a deeper level.

Hypnosis induces a trance-like state where the conscious mind becomes more receptive to suggestions, allowing for direct communication with the subconscious.

NLP combines language and behavioral patterns to reframe and transform limiting beliefs and patterns stored in the subconscious.

As you read about the different methods in this book, keep in mind that some of these can work simultaneously, but with others, it'll be much more effective if you pick one method, at least to start.

You can better focus your intention by giving a couple of them your full attention.

As you gain experience and confidence, you can incorporate additional techniques into your practice. By expanding your range, you tap into different aspects of the subconscious mind and amplify the reprogramming process. Experimenting with various techniques and finding a combination that works best for you can further enhance your ability to take control of your subconscious mind.

And keep in mind that you can always incorporate more of these techniques as you practice a little bit at a time.

Chapter 6

The Power of Affirmations

"You will be a failure, until you impress the subconscious with the conviction you are a success. This is done by making an affirmation which 'clicks.'" - **Florence Scovel Shinn**

―᷾᷾―

Everyone is familiar with affirmations. These are positive statements that you repeat to yourself each day, aimed at rewiring your subconscious mind and instilling empowering beliefs.

Affirmations can be powerful tools for personal transformation, but to harness their full potential, it's crucial to use them correctly. Let's review how.

Basic Rules for Using Affirmations

Avoid affirmations that are clearly untrue. Your subconscious mind is flexible, but it's not willing to consider the ridiculous seriously.

Repeating, "I am a world-renowned expert in astrophysics," is factually untrue if you're just starting your journey in that field. It's also too big a leap for your subconscious mind to make.

While it's great to aspire to such achievements, it's more effective to create affirmations that bridge the gap between your current state and the desired outcome.

Instead, you can use affirmations such as, "I am dedicated to expanding my knowledge in astrophysics and becoming a respected authority in the field."

This statement acknowledges your current commitment to your aspiration while setting an intention for growth and expertise. By focusing on the process and the steps you need to take, you create a more believable and actionable affirmation that your subconscious mind can embrace.

Once you've aligned your affirmation with reality, you must make it powerful. You do this by using imagery and emotions to sell it to yourself.

See yourself as thin, financially successful, or selling your first successful novel. Use these mental images to generate the same emotions you would experience in that situation. We'll look at this in more detail a little later.

Finally, repeat your affirmations three sessions per day for 5-10 minutes each. A good schedule is right before falling asleep, right after awakening, and any opportune time during the day. For example, if you want to strengthen your confidence, repeat the affirmations when you're feeling self-doubt.

Why Affirmations?

Life brings a lot of negative thoughts into our lives. The challenging experiences we face, the mistakes we make, and the uncertainty of the future can all contribute to your negative thoughts and emotions.

The trick is in learning how to push those negative thoughts out of your mind and replace them with something positive. This is how you can use the power of positive affirmations.

As you use them, remember, affirmations are not mere wishful thinking. They are powerful tools for self-empowerment and personal growth.

Used correctly, they can help rewire your mind, dissolve self-limiting beliefs, and align your thoughts and emotions with your desired reality. By integrating affirmations into your daily practice with positivity, specificity, consistency, visualization, and belief, you can unleash their full potential.

This is how you can use them to manifest positive changes in your life.

Creating Positive Affirmations

A positive affirmation is a simple yet powerful statement that you say about yourself. It serves as a tool to change the way you habitually talk to yourself and transform the inner dialogue that shapes your self-perception. Consciously using positive affirmations can reverse the negative images and self-talk that have been ingrained in your mind.

When you consistently tell yourself affirmations like, "I totally and unconditionally accept and love myself, just the way I am" or "I love my life and am excited about my future," you are intentionally replacing your stored negative thoughts and images with uplifting and empowering ones.

These positive affirmations serve to counterbalance the self-criticism and self-doubt that you've been telling yourself. They also help shift your focus towards the person you aspire to become.

By affirming self-acceptance, self-love, and enthusiasm for the future, you create a mental environment that nurtures growth and self-improvement. These empowering thoughts gradually reshape your self-image, enhancing your confidence and resilience.

To create a daily positive affirmation, write out something you want in your life or a quality you want to possess.

Make a list of these affirmations. Write them in the present tense, using positive language.

This can be something like, "Every day I am improving my health through diet and exercise."

Now start reading these statements to yourself every morning and every night.

Used this way, affirmations enable you to overwrite negative self-talk, cultivate self-acceptance, and redirect your focus toward a more positive and empowering mindset. By consistently practicing positive affirmations, you build a foundation of self-love, self-acceptance, and optimism that support your personal growth and helps foster a more fulfilling life.

Maximizing Your Affirmations with the 4 P's

When you create affirmations, remember the four P's:

1. Make it Personal.

Your affirmation must be about "me" or "I." Affirmations involving others seldom, if ever, work.

2. Make it Positive.

Your affirmation needs to be about something that moves you toward your goals.

For example, "I successfully complete all my courses with a score of 88 percent or better" instead of "I don't fail any of my courses."

3. Make it in the Present

"I am earning $100,000 a year" instead of "I'm going to earn $100,000 a year."

4. Make it Powerful!

After using the three P's making it personal, positive and in the present to form your affirmations, use the final P to make it powerful!

As you say the affirmation aloud, bring up all up the corresponding feelings attached to it.

Saying, "I am wealthy" while feeling poor only sends conflicting messages to your subconscious!

This never works!

It's important to keep in mind that, to make your affirmations more powerful, you need to strive to feel the corresponding emotions. This will help your subconscious believe them.

Now, repeat, repeat, and repeat.

Affirmations won't work if you say them just once or twice. Recite them many times throughout the day for the best results.

The great thing about using affirmations is that you can fit them seamlessly into your routine.

For example, if you are using affirmations to counteract stress, when you start to feel stressed, you can repeat something like, "I let go of unnecessary stress in my life to make way for peace."

You can also use, "I deserve peace and recognize its value for my health, productivity and relationships."

Practicing Makes Everything Better

Don't just create one or two daily positive affirmations. Create as many daily affirmations as you can think of and repeat them to yourself every morning and every night.

As you make this a habit, you'll begin to notice your affirmations are taking effect and you'll start feeling differently about yourself.

Another benefit is, as you use your positive daily affirmations habitually, you'll find yourself using more positive speech in general. This is even when you're not talking in reference to yourself.

How much help you'll get from your positive daily affirmations will depend on the commitment you make to them. A rule of thumb says it takes 30 days to develop a habit. So, commit to practice for at least 30 days. The key is to practice every day, so it becomes second nature to you.

As you develop the habit of thinking positively, you will find negative events are less likely to linger and take root in your mind. You are successfully reprogramming your subconscious mind!

Chapter 7

Visualization

"If you don't have a vision for the future, then your future is threatened to be a repeat of the past." -
A.R. Bernard

People around the world use visualization to inspire and motivate them. For example, Bruce Lee, the legendary martial artist, and actor, was a firm believer in the power of visualization.

Jim Carrey, the actor and comedian used visualization to help him stay focused on his goals and imagine himself achieving success.

Tiger Woods says that he uses visualization to help him plan his shots and imagine the ball's flight path before he hits it.

Visualization is more than just thinking about what you want. Visualization is an effective tool that harnesses the power of your subconscious mind to help you achieve the things that you want in life.

Your mind has an amazing ability. It can help you create your own reality. You can do this by using visualization to create the reality you desire.

Our thoughts, like everything, are made up of energy. This energy sends vibrations out into the universe.

As you send out the vibrations, they attract similar energetic vibrations and draw them to you. The more positive energy you put out, the more positive energy you'll receive.

This universal law of attraction is at work in your life, whether you believe in it or not. By understanding how these universal laws work, you can make a conscious choice to work with them and manifest significant changes in your world.

I Don't Think I Can Visualize

Many people say they aren't good at visualizing, but in reality, most people are fairly visual. What this means is that there is a very good chance your subconscious mind will respond well to pictures.

This is what visualization is, making pictures in your mind.

If you're not yet aware, I can assure you that you are already using some sort of visualization. For example, if you "see" that a meeting with a person will not work out and it doesn't, you are using visualization.

This type of visualization is called negative visualization. You don't want to develop this type of visualization. You want to develop the ability to visualize positively or positive visualization. Positive visualization is focusing on the good things you want to happen in your life, and you want to do this on purpose.

To do this successfully, you will need to set aside a certain amount of time, let's say 10-15 minutes a day, visualizing the positive life events you want to experience.

Positive Uses of Visualization

Here is a short list of some of the things you may want to visualize:

- A great relationship with your family.

- An abundance of money.

- A beautiful home.

- A fulfilling spiritual life.

- Close and meaningful personal relationships.

- The type of car you've always dreamed of.

- Work you can be passionate about.

- Vacations to places you've always wanted to visit.

- A healthy, slender, fit body.

- Anything else you wish to draw into your life.

As you make a habit of visualizing what you want, you'll find yourself redrawing the negative pictures of experiences you've stored in your memory. You can turn things like fears, worries, and doubts into the things you actually want in your life.

The Visualization Process

Hopefully, as you've gone through this book, you've become open to trying something new. You'll find that with a little practice, visualization isn't complicated at all. Right now, let's look at the basics.

The actual act of visualization goes something like this:

Start with the Goal

Ask yourself, "What do I want to achieve?"

This should be a specific milestone in your personal or professional life. Some examples are starting a successful business, running a marathon, writing a book, or achieving a state of inner peace.

Once you figure this out, everything else will fall into place naturally.

We'll explore some tips for choosing goals in a moment.

For now, we'll assume you have figured out what you want.

Imagine the Goal

This isn't some easy low-level image you're creating in your mind. You want more than an image of your boss telling you that you've just received a promotion.

You want to picture your goal in the most intense, realistic detail you can imagine. As you view the image, layer in its sights, sounds, smells, tastes, and tactile sensations.

Create the emotions which go along with it.

Experience how reaching this goal is going to make you feel. Explore any other feelings that come up as you imagine your goal.

Revisit the Goal Regularly

You don't use visualization only one time and achieve your goal. You need to take it out and go over it again and again.

Make a commitment to give yourself visualization time daily to revisit the images you've created. Remind yourself of them throughout the day.

Accept the Reality of the Goal Completed

This step takes a little leap of faith.

You are reminding yourself that you're already there. This goal is already achieved. To make it a reality, you simply must keep moving toward it.

Here's something else to keep in mind. Once you've created your visualization, you need to remember to disengage from it.

This is an essential step in the process. Otherwise, you might get caught up in living the vision to the point where you forget to translate the vision into the steps it takes to make it a reality in your life.

Visualization sets up your future success. You're still going to have to do the work to make that success a reality. This comes about by accepting the opportunities that come your way in the wake of this visualization.

Here are some further things to think about, which might help with this process:

Home in on What You Want

If you're having trouble visualizing your goal, it might be because your goal isn't specific enough. Take a step back and look at your goal objectively.

Is there any particular part of what you want to do which intrigues you more than other parts?

If you find you're having trouble finding your goal, think about the things you're passionate about. Make a list of them.

What is a common thread you find recurring in this list?

Experiment with Different Techniques

We're all different. This means it should not surprise you to find not every method of visualization works the same for every person.

For example, if you find yourself having trouble thinking in pictures, try using words or feelings to convey your intention.

Maybe you need to write down the visualization first. You can also give yourself a 'script' to read aloud to yourself.

For some people, making a collage of pictures, or vision board, to look at as you visualize your goals helps.

Feel free to explore different options to find what works for you. As long as the result is the same, use what works to help you visualize your goal.

Meditate

If you're having trouble quieting your mind, practicing meditation will help to calm it. By learning how to stay in the moment and calm your breathing and your thoughts, you'll better prepare yourself for visualization.

For more on meditation, see chapter 8.

Narrow Your Focus

If you're having trouble concentrating, start smaller.

Visualizing for only five or ten minutes to begin with will teach you how to focus in small doses. Keep in mind that, in the beginning, you don't need to worry about a complex visualization.

As you're beginning, keep things simple. As you become more comfortable with this level of concentration, you can expand the time you spend visualizing.

You can also include more details.

Silence the Inner Voices

It's hard to focus when you're listening to a hundred different thoughts all clamoring for attention.

You can use this short visualization exercise to block them out.

Imagine you're in a room with all those voices outside of an open window. Now shut this nice thick double-paned glass window and enjoy the quiet within.

Note: You can also use the 5 R process in chapter 4 to help you silence those inner voices.

Trust the Process

The more you question whether things will work, or hold back on accepting what you're visualizing, the less likely you will experience success.

Take a leap of faith. Trust the process.

Use Repetition

Visualization never works well if you're not putting in the time. Visualization needs to be part of your daily routine.

For the best results, schedule in a regular visualization time. This might be right when you get up or before going to bed at night. You can even try both if you like!

Create a Relaxing Atmosphere

As mentioned before, visualization works best with a calm mind. This is why it is crucial to pay attention to your surroundings.

How can you minimize distraction or noise?

You could create a calm mood or ambiance by using candles, soft music, or dimmed lights.

Do what you need to do to create a calm port in the stormy seas of life.

Use the Internet for Help

When you're just starting out, this whole visualization process might seem like a steep, uphill climb.

To help, try one of the many videos offered online. There are also podcasts that offer guided visualization. Once you learn how to relax into the visualization, it's simple to begin creating your own.

Get Help

Do you know someone who practices visualizing regularly? Find a mentor to answer your questions, or help you create visualizations that are suited for your needs.

Be Open to What Happens

Once you've finished visualizing, be open to what happens. You'll find that things will come up in your life, some guiding you on a particular path.

Accept where they lead.

As new opportunities present themselves, embrace the adventure of learning to move with confidence toward your dreams.

Perseverance is Key!

Nothing happens overnight. This is especially true when you're just starting and learning a brand-new way of thinking.

The key is to keep at it. Give yourself time to become accustomed to how visualization works.

How to Put Visualization to Use

Follow these steps to use visualization effectively:

1. Find a place that's quiet. An environment without distractions.

2. Close your eyes. Slowly breathe in through your nose to a count of five, then exhale slowly through your mouth, again, to a count of five or more. As you breathe in, your diaphragm, or stomach, should expand. As you exhale, it should flatten.

Take a minimum of three to five breaths like this to slow your heart rate and calm your mind before you begin.

3. In your mind, create a clear mental image of what you desire. Think about the details. Notice the colors, textures, sounds and smells associated with your vision. Imagine touching the things in your vision and tasting the air.

Let this image play like a movie in your mind. See yourself as the director and leading character.

Picture yourself already doing or having the things you want to manifest. See yourself as you accomplish what you want to do almost effortlessly.

4. Set aside two or three times a day to practice visualizing the things you want to achieve. To successfully use this or any tool in this book, you need to consistently practice every day.

Use visualization to respond to negative thoughts.

If you start to experience negative thoughts during your day, you can use visualization to stop them. Take a few moments to visualize positive thoughts you can use to replace your negative thoughts.

Follow each step of this process through to completion. Let yourself feel the emotions connected with achieving your goal.

Try to remove any feelings of doubt about its achievement. Visualize the outcome as though you've already finished it.

As you go through your visualization, take your time. Give yourself the opportunity to experience the vision fully. Around ten minutes or so is usually sufficient.

You can use visualization to help you clarify your desires. Keep in mind that you may not receive the answers you're seeking in the way you expect. Be open-minded to different manifestations of what you're trying to achieve and don't expect magic.

For example, if you want an increase in your income, don't expect to win the lottery, or receive a check from a long-lost relative. Instead, be open to a new job, a promotion, or some other opportunity to earn extra income.

Sometimes things happen quickly. Other times what you want may come to pass gradually. When you keep an open mind, more opportunities can present themselves to you.

Let's get you started with some visualizations you can try immediately to achieve some of the more common goals.

Visualization Exercises

There are many ways you can use visualization in your life. Once you get the hang of it, you'll be amazed at all the things you can accomplish with the aid of your imagination. Let's begin with some super simple visualizations.

Simple Visualizations for Beginners

Life of the Party

This is a great visualization if you're feeling a little socially anxious.

Let's say you have been invited to a specific social function. Begin your visualization by asking yourself what kind of person you want to be by the end of the event.

Do you want to be really social? A little more gregarious?

With your goal in mind, picture yourself.

Next, picture the event. Watch as you move inside this event, easily interacting with people.

Focus on the positive responses you get and the emotions you feel. Search for happiness points.

Combine this with the self-talk you need to support this version of you. This can be things like "I am comfortable here," and "I like talking to people."

Take your time and practice the scenario repeatedly. By the time you reach the event, the interactions you encounter should feel familiar and natural.

Safe Place

There are places in our lives where we feel totally safe and secure. If you find yourself feeling anxious, use this visualization to bring back those memories.

Start by visualizing a place where you feel safe and secure. Use as much detail as you can manage.

Make sure to use all your senses. Now see yourself in this place. Let the safety and comfort of this place surround you until you feel peaceful and calm.

Balloon

This is a great visualization to get rid of troubling thoughts.

Use the deep breathing we talked about earlier to put yourself into a calm and relaxed state of mind. As you begin to feel safer and more secure, allow the bad thought to appear in the space next to you.

Look at it and examine it.

Identify this for what it is – simply a troubling thought. Do not give it any more power than this.

Now, imagine a balloon surrounding the thought until it's entirely encased and bobbing on the end of a string you hold firmly in your hand. As you watch the balloon, decide to let it go and watch it float away into the sky until it completely disappears.

Practice Scenarios

Whenever you have something new coming up that worries you, such as a speech or some other event where you're going to be called upon to perform, use visualization to practice your performance in advance.

You already have your goal in mind. It is the action you will be called upon to complete.

As thoroughly as you can, create the venue where this event is going to take place. If possible, visit the place in advance. You can use this experience to provide you with accurate, in-depth detail.

Next, imagine yourself there, performing your activity. See yourself from the audience's viewpoint, watching as you perform.

Then, run it through again, looking at this event through your own eyes. Experience every emotion that comes with this event.

Run through it from start to finish. Make sure you end it on a triumphant note.

Lemon to Lemonade

In this visualization, you need to find an actual piece of fruit. A lemon works well for two reasons, its simple shape and pleasant aroma.

Start the visualization by calming yourself using your breathing. Relax entirely and thoroughly.

As you reach the state where you still have control of your body but are feeling deeply immersed in the visualization state, open your eyes to study the lemon.

Smell it.

Touch it.

Use all your senses to take it all in.

Close your eyes. Recreate the lemon in your mind.

Make sure you use the same level of detail. You can use this exercise to get used to visualizing actual items in such a way that makes you feel comfortable manipulating them. It also is a model for imagining something you already have.

Practice this until you can recreate the lemon perfectly in your mind.

After you've mastered this visualization, you're going to shift your point of interest onto something you desire.

For example, if you want more money, hold out a hundred-dollar bill and use it in the same way. Learn how to construct the money accurately in your mind.

If you want something else, find an image you can use to help you visualize. If this is moving to the house of your dreams, use a picture of the kind of home you desire.

Whatever you pick, hold the image in front of you, memorizing every detail of it. Once you've finished, recreate the image in your mind with as much detail as possible. This is how you can use visualization to get the things you want.

Use Your Senses to Enhance Your Visualizing Experience

"You must watch the pictures that you paint with your imagination. Your environment and the conditions of your life at any given time are the direct result of your own inner expectations. If you imagine dire circumstances, ill health [such as cancer] or desperate loneliness, these will be 'automatically' materialized, for these thoughts themselves bring about the conditions that will give them a reality in physical terms.

If you would have good health, then you must imagine this as vividly as you fearfully imagine ill health." - **Martin Brofman, Ph.D.** (Cured himself of terminal cancer)

In NLP, or Neuro-Linguistic Programming (we'll talk about this in chapter 11), we use the power of the senses to boost the intensity of visualization even further.

By including the feelings, sights and sounds of what you want to accomplish you produce strong, positive emotions. Your subconscious mind then associates the strong feelings with the accomplishment of your deepest desires.

Because your subconscious mind absorbs these messages as if they're real, the strong feelings associated with them will motivate you to take the action you need in the real world to achieve your goals.

This is the true beauty of visualization. By seeing your goals as if they were already accomplished, it makes it easier to undertake the action needed to complete what you've set about to do. It gives your subconscious mind the power to bypass any limiting past messages and focuses it on the images of what you truly desire.

Chapter 8

Meditation

"Now I meditate twice a day for half an hour. In meditation, I can let go of everything. I'm not Hugh Jackman. I'm not a dad. I'm not a husband. I'm just dipping into that powerful source that creates everything. I take a little bath in it." - **Hugh Jackman**

MEDITATION IS A TRANSFORMATIVE practice that has stood the test of time. It is rooted in ancient traditions in India and China that go back thousands of years.

The goal of meditation is to help you focus your attention and awareness. This helps you achieve a state of mental clarity and emotional balance.

Think of it as a form of mind training that can help you adopt a more positive outlook and cultivate inner peace. This could be by letting go of the past or by stopping dwelling on the future. It could also be by simply bringing a more positive mindset into the world.

Meditating helps you gain a deeper understanding of your thoughts, feelings, and reactions. By practicing consistently, it can help you develop a profound sense of inner calm.

The Purpose of Meditation

"Meditation is a vital practice to access conscious contact with your highest self." - **Wayne Dyer**

The purpose of meditation is to clear your mind of the constant barrage of negative distractions we all experience. These are things like noisy neighbors, bossy officemates, that parking ticket and unwanted spam.

Meditation helps you to relax your mind so you can allow these distractions to float away gently. This leaves you feeling refreshed and focused.

The practice of meditation allows you to learn how to relax your mind and your body. It can help you develop your life force, also known as Qi, Chi, Ki, and prana.

At the same time, you can learn how to love yourself and others again, have compassion, and improve your generosity. It can also help you forgive yourself and others.

The goal is to relax your mind and focus your awareness on your thoughts and feelings. You do this by clearing your mind and thinking only about one thing at a time. You may wish to focus on things like losing weight or forgiving someone for a recent action they took.

Practicing Meditation

Nowadays, there are many ways to practice meditation. Some people find tranquility by practicing yoga. Others prefer the fluid movements of Tai Chi or Qigong.

If you're more of an audible learner, repeating phrases or quotes might be your go-to. Some people even meditate while walking.

Gone are the days of strict guidelines and expectations on how to meditate. Now it has more to do with your personal preference.

You can choose a peaceful spot where you won't be disturbed. But if you can't find a peaceful oasis, don't worry, you can still meditate anywhere.

Your living room, a stroll around the block, or even your office can become your very own meditation space. All you have to do is close your eyes and breathe deeply for a few minutes to get your Zen on. It doesn't matter where you are.

You Don't Have to Be a Contortionist

Most people think that they have to mediate in the same positions they see on television. You've seen them. The ones with impossibly arched backs and painful-looking contortions.

You don't have to meditate this way.

What you need is to be in a comfortable position that is conducive to relaxation and concentration. You can meditate while sitting cross-legged, standing, lying down, and, as mentioned before, even walking.

The trick is to find a position that allows you to relax and focus. This can be sitting or standing, with your back straight, but not tense or tight. The same goes for any position you decide to use. The only no-no's are slouching and falling asleep.

Chanting

Have you ever watched monks on National Geographic or Discovery making repetitive sounds? Those sounds are called mantras.

Basically, a mantra is a short phrase or simple sound that holds some deep spiritual significance for those who practice it. But don't worry, you don't need to chant a mantra to reap the benefits of meditating.

Most people find it easier to focus on things like slow breathing or humming to get into the meditation zone. If it works for you, use it.

Meditation Fundamentals

The fundamental principles of meditation remain the same no matter what approach you choose.

Of all the principles, the most important is that of removing obstructive, negative, and wandering thoughts and fantasies. You want to relax the mind and gain a deep sense of calm focus. This process clears the mind of life's negative debris and prepares it for a deeper quality of activity.

Some practitioners choose to shut out all sensory input. They eliminate all sights, sounds, and leave nothing to touch.

These people believe that this allows them to detach from the commotion in the world around them.

This can work especially well for beginners, although the lack of stimulus may seem deafening at first. All of us are too accustomed to constantly hearing and seeing things that prevent us from calming and focusing our minds.

When you shut out those distractions, your mind will search for them, but as you continue, you will find yourself becoming more aware of everything around you.

Basic Meditation

There is no absolute right or wrong about meditating. If you can, choose a place that has a soothing atmosphere. This can be your living room or bedroom. Any place that you feel comfortable will do. You may want to arrange the space so that it is soothing to your senses.

Adopt a comfortable position. This can be sitting down in a chair or lying back on your bed. The important part is that you are comfortable. This allows you to focus entirely on your breathing.

Silence helps most people relax and meditate, so you may want a quiet, isolated area far from your ringing phone or the humming of the washing machine.

For some people, pleasing scents also help. If this is you, stocking up on aromatic candles can help.

Most people meditate with their eyes closed, simply focusing on their breathing and counting their breaths. The basic principle here is to focus.

Another method is focusing on a certain object, such as a lit candle, while keeping your eyes open or on a single thought or affirmation.

Other Forms of Meditation

The most important thing about meditation is making it a priority in your life. Whether you choose to do a guided meditation or simply focus on your breath, it's important to carve out time in your day for this practice.

It's easy to brush it off as something you'll do when you have some free time, but if you don't make time for it, it won't happen. So, make meditation a priority in your schedule.

No matter if it's the first thing in the morning or right before you go to bed at night, make a commitment and meditate.

Now, let's look at some other types of meditation.

1. Mantra-based Meditation

This form of meditation uses a specific phrase or sound. As described before, these sounds are known as a "mantra." While some forms focus primarily on one's breathing and bodily sensations, mantra meditation focuses on a repeated sound, chant, or phrase.

An example of a mantra is, "May my heart be kind, my mind fierce and my spirit brave."

You can search the internet for one that fits you.

2. Visualization Meditation

Visualization meditation focuses on images to achieve a specific objective. For example, athletes use visualization meditation to improve their performance.

With visualization, your focus is on your desired outcome. You imagine a detailed picture of what the outcome looks like, feels like, and smells, sounds, or tastes like.

3. Moving Meditation

Some people prefer this type of meditation because it combines exercise, such as walking, with meditation. Moving meditation is actually a yoga practice combined with mindfulness.

Sometimes called "Breath walk," you synchronize your breath with your footsteps while focusing on bodily awareness.

Other examples are practicing Tai Chi and Qigong.

4. Body-Scanning Meditation

This meditation involves bringing awareness first to one specific part of the body and then moving on to others. Thus, the term body scanning.

For example, you focus on the physical sensations, stress, or tension in your left arm and then intentionally release all tension in that area. Repeat with your right arm. Now your left leg.

Go through all the other areas of the body until fully relaxed.

5. Transcendental Meditation

If you're looking for a way to meditate without having to join a cult or travel to a mountaintop, then Transcendental meditation might be just what you're looking for.

Though often associated with certain organizations, this refers to the simple method. And don't worry, you don't have to be a bohemian or Buddhist monk to get into it.

People of all ages, backgrounds, and beliefs practice this natural technique. In fact, about 5 million people worldwide do it across all age groups, cultures, and religions.

Due to its popularity, there have been over 350 research studies done on transcendental meditation. Even Stanford and Harvard medical schools have conducted studies.

While this method may be new to some people, its root comes from an ancient Indian (the continent) method or tradition known as Vedic. This tradition was handed down over many millenniums. Maharishi Mahesh Yogi first introduced it to the western world about 50 years ago.

The instructors that teach this method today still use the same procedures and principles that have been used for thousands of years.

This method is very simple to follow because it doesn't require a lot of concentration, as other meditation methods tend to do. The only effort that is required is for you to find a comfortable sitting position in a chair.

Transcendental meditation is the process of transcending into a state of natural awareness. While your body becomes restful, your mind remains alert. This state is so natural that people find it extremely easy to use this process.

If you have never tried any form of meditation, and would like to begin, transcendental meditation can be a great starting point.

6. Several Other Types of Meditation

There are many mindfulness meditations you can practice when you focus on your breathing. These include mindfulness listening, mindful immersion, mindful observation, and mindful awareness.

These other forms of mindfulness meditation involve setting your intention on a specific thing or task and paying attention to the details.

For example, mindful immersion is a form of mindfulness where you immerse yourself completely into the task at hand. This creates an entirely new, unique, and fresh experience.

We'll examine mindfulness later in this book.

As you set out to create a regular routine for practicing meditation, you might want to mix up the styles. This is especially useful if you feel your practice is monotonous or you're becoming bored.

An example might be practicing mindful listening the first week then choosing to practice moving meditation in the second week.

Chapter 9

Hypnosis

"Hypnosis is a normal and natural way of knowing your inner self and augmenting it with virtues like self-belief." - **Dr. Prem Jagyasi**

HYPNOSIS IS AN ANCIENT practice that has stood the test of time. Its rich history spans across many cultures and societies. It has benefited millions of individuals.

People around the world have used hypnosis for self-improvement since ancient times. Its effectiveness has been well-documented, making it an established practice accepted and used worldwide.

What Is Hypnosis?

"You use hypnosis not as a cure but as a means of establishing a favorable climate in which to learn." - **Milton H. Erickson**

The word "Hypnos" comes from the Greek to "to sleep." But hypnosis is, in fact, very different from sleep.

Hypnosis is a state of mind characterized by relaxed brain waves and hyper-suggestibility. It can be a trance state characterized by extreme suggestibility, relaxation, and heightened imagination.

It's not really like sleep because your mind is alert the whole time. People who have experienced hypnosis compare it to daydreaming, or the feeling of "losing yourself" in a book or movie.

While hypnotized, you are fully conscious, but you tune out most of your surrounding stimuli. You focus intently on the subject at hand, to the near exclusion of any other thought.

By itself, the hypnotic state is only useful for the relaxation it produces. The real importance of hypnosis is that while you are in the hypnotic state, your mind is open and receptive to suggestions for healing and emotional change processes.

How Does Hypnosis Work?

"Want to lose weight? Kick a bad habit? Well, you might want to try hypnosis! ... no longer regarded as mere hocus-pocus, it's been shown as an effective means of helping people quit smoking, shed pounds, reduce stress, and end phobias." - **Jane Pauley**

Hypnosis is a perfectly normal state that just about everyone has experienced. In fact, The World Health Organization reports that 90% of the general population can be hypnotized.

What we call "highway hypnosis" or being hypnotized by the road as you drive is a natural hypnotic state. You drive somewhere and don't remember driving or even remember seeing the usual landmarks. Your mind has been driving your vehicle on automatic pilot.

This same, natural hypnotic state also exists when you become so involved in a book, movie, TV show, or any other activity that you block out everything else. Someone can come into where you are to talk to you, and you don't even notice them.

These are natural hypnotic states. Whenever you concentrate that deeply, you automatically slip into this natural hypnotic state.

Hypnosis works because it allows you to tap into the part of the mind that this book is all about – your subconscious!

As you know, the subconscious mind is actually your control center. The power of hypnosis is it can save you a lot of time by letting you speak directly to it.

Hypnosis works by putting your mind into a deeply relaxed state. This allows positive and healing suggestions to sink deeply and quickly into your subconscious mind.

I say positive suggestions because all research has demonstrated that hypnosis cannot force you to act against your morals and ethics.

Hypnosis Facts

"All problems in life are problem trances, and all solutions are solution trances." - **Igor Ledochowski**

Here are few facts about hypnosis:

Psychologists use hypnosis to calm and soothe people who are in a state of trauma or are nervous.

Medical practitioners use hypnosis as anesthesia. Surgeons use it during surgery, dentists during tooth procedures, and doctors and midwives during childbirth.

Another use of hypnosis is for post-surgery or post-operation to prevent infection or unpleasant side effects.

Mental health professionals use hypnosis as part of the therapy for patients with psychological conditions.

Hypnosis is used to help both patient and doctor understand health and mental conditions, the causes and possible course of action to be taken.

People use hypnosis to help them curb problem habits such as smoking, eating disorders, or unpleasant behavior.

Hypnosis is effective in dealing with psychosomatic problems or physical illness that can be rooted in a psychological condition.

Learning ability, physical performance and social attitudes can also be improved through hypnosis.

Forensic hypnosis is used by the legal system to aid in criminal investigations.

The Effect of Hypnosis on Your Subconscious Mind

"I should have done it years ago. It's amazing I didn't even want cigarettes anymore." - **Matt Damon**

As explained before, all your habitual behaviors reside in your subconscious mind. They named it the subconscious because it is deeper than your conscious mind.

It's below your level of consciousness.

Imagine that there is a trapdoor between your conscious mind and your subconscious mind.

Normally, the trapdoor is closed. It opens when your brain waves slow down to a relaxed, alpha brain wave level.

This is exactly what happens when you are asleep. The door opens for short periods, allowing ideas, images, and thoughts to come out of your subconscious mind. We call what comes out in your sleep "dreams."

When you are in a state of hypnosis, the door also opens. This allows helpful suggestions to go into your subconscious mind.

Some people have found this useful for recovering forgotten thoughts and memories.

Three Keys for the Successful Use of Hypnosis

The keys to successful hypnosis are self-motivation, repetition and believable suggestions.

1. All motivation to change must come from within.
You already know this.

No matter how much you know someone else needs to change and no matter how much you want him or her to change, you cannot make him or her change.

The same goes for you.

All motivation for change must come from inside you.

For example, let's say you are trying to change because someone else wants you to "lose weight" or "stop smoking."

This just doesn't work.

I've worked with many people for weight loss or to quit smoking who came to me because their physician or spouse wanted them to change. These people do not respond well, if at all, to hypnosis.

On the other hand, someone who has a goal of quitting smoking or losing weight often responds quickly and easily.

This means, before you start to use hypnosis for a self-improvement program, you need to get clear on what you want to change and why you want the change to happen. This clear intention to change will help the hypnotic suggestions you're given to take hold and manifest themselves in your everyday life.

If you are having difficulty figuring this out, find a qualified hypnotherapist to help you move through this step more quickly.

2. Repeat, repeat and repeat some more.

You need repetition to succeed in almost everything you undertake.

For example, if you wanted to improve your fitness level you wouldn't expect to go to the gym once and have the fitness level you desire. You have to go to the gym regularly.

The same goes for hypnosis. To use it successfully, it will take regular repetition, at least while you are developing your new and empowering habits.

Our behaviors, for the most part, are just habits. They are things we've learned to do automatically over time.

The interesting thing is that most behaviors are neither "on" nor "off" but happen or are "triggered" due to different types of stimulation.

Sometimes those triggers are stimulated by very different events or situations.

This means that you may come to a hypnotist one day with one set of triggers to work on and then feel fine when you leave, but the next day, after being exposed to a different set of triggers, you experience the same behavior again.

The good news is you got one set of triggers under control.

Now, in order to modify your habitual patterns, you have to address the other triggers.

This can take a few sessions to get ironed out. This is why many hypnotists give you self-hypnosis recordings to listen to at home.

3. The suggestions must be realistic or believable.

The third key for the successful use of hypnosis for personal change is the suggestions you are given must be realistic or believable.

As we discussed in the section on affirmations, your mind will not accept a suggestion that you do not or cannot believe.

No matter how positive it is, or how much good it will do for you, to accept a suggestion, your mind must first accept it as a realistic possibility.

For example, telling a chocoholic that chocolate will be disgusting to them and will make them sick is too big a stretch for their imagination to grasp.

If a suggestion like this even took hold, it would only last a very short time because it is so unbelievable to a real chocolate lover.

In cases like this, one of the successful weight loss suggestions I have used is that the next time the individual eats chocolate it will not taste quite as good as the time before. This is far more realistic and believable and therefore acceptable to most people. Then, with enough repetition (remember number 2) over time, chocolate loses much of its allure and eventually lessens its control over that person.

Is Hypnosis Dangerous?

Many people have been told by people with certain worldviews or religious beliefs that hypnosis can be harmful. They say that the hypnotist can take over your mind and make you do all kinds of immoral actions.

This is 100% totally untrue.

The facts are HYPNOSIS IS NOT DANGEROUS. There are almost no risks when used by trained professionals. Even amateurs cannot cause you to harm yourself or anyone else.

As mentioned before, hypnosis cannot make you do anything that is against your moral or ethical values.

Chapter 10

Self- Hypnosis

"Your self-image is the result of all you have given your subconscious mind as a database, so regardless of your background, what you are willing to become is the only reality that counts." - **Paul J. Meyer**

Self-hypnosis is also an excellent method used by people to help them sleep, reduce stress, lose weight or to eliminate unhealthy habits. Because it only involves one person, it is the most commonly used form of hypnosis.

What Is Self-Hypnosis?

Self-hypnosis is exactly what the term suggests, hypnosis that you do to yourself. You can experience self-hypnosis by putting yourself in a trance while giving yourself suggestions or using a recording made by you or someone else.

You can use this process to promote positive change or achieve specific goals.

The Risk of Self-Hypnosis

One of the biggest risks of self-hypnosis involves falling asleep.

I have worked with thousands of hypnosis clients. Even in a clinical setting, clients fall asleep during a hypnosis session. This is usually because they came in tired or became too relaxed.

If you are tired or become too relaxed, you may move from the state of hypnosis to the normal sleep state. While this is fine if you were planning to go to sleep right after the trance, if you have other plans after listening to a hypnosis recording, you may want to set an alarm to wake you at the end of the session just in case you fall asleep.

In relation to this, never listen to a hypnosis recording while driving or doing anything that requires your entire attention. This is because, if you are tired or become too relaxed, you could fall asleep. This is very dangerous for you and everyone else on the road.

Don't even listen to a self-hypnosis recording if you are a passenger and there is a chance the driver can hear it. You don't want to take the chance that the driver might fall asleep.

You may feel a little bit drowsy immediately after you go through a self-hypnosis exercise. This isn't anything to worry about. Feeling a little sleepy is normal and will go away shortly, nearly always within a few minutes.

However, don't attempt to drive or do anything that takes a lot of concertation until you feel your complete orientation has returned.

Preparing for Self-Hypnosis

First, find a quiet place where you can relax and are unlikely to be disturbed. Sit or lie down. Your goal is to relax comfortably.

Once you are in a relaxed position, choose a "trigger word" to help you to relax. Good trigger words are "let go" or "relax." You are going to repeat your trigger word at the beginning of each breath you take.

Next, choose an affirmation or positive suggestion to repeat. For example, if you feel stressed and want to relax, try using something like, "I am feeling relaxed and comfortable."

Finally, try to visualize a relaxing scene. For some people this is the beach. Others like to imagine walking or sitting in a meadow or garden. My personal favorite is imagining being in an Asian garden with lots of greenery. Whatever works is perfect for you.

If you are having problems visualizing a scene at first, simply try to imagine a soothing, shapeless cloud of mist surrounding and holding you gently. You can also try imagining resting on a cloud or floating down a gently flowing river. Again, it doesn't matter what it is as long as it works for you.

The Steps to Hypnotizing Yourself

1. Sit comfortably on a couch or a chair.

Sit in an upright position. Close your eyes. Let your arms relax at your side or in your lap. As you sit there, just relax.

2. Clear your mind of all thoughts.

How do you do this? A simple method is to imagine a big dumpster or a very large metal box that's on a large movie screen in your mind. Watch the movie screen as the box or dumpster opens. Now, put all your thoughts and worries in it one by one.

Put your bills... your love life... your children... your job... any and all your problems you can think of. After you've put them all in, close the lid and lock it! Finally, imagine pushing the dumpster off the edge of your mental movie screen.

3. Sit quietly for a minute.

As you ready yourself to start, remind yourself that each part of your body is becoming more and more relaxed as you progress through your self-hypnosis session.

This reminder is vital, and you need to give it to yourself. This is your trigger to begin relaxing your body and mind.

4. Now begin.

Starting with your head, imagine that every part of your head is relaxing. Your scalp, your eyebrows, your eyes, your cheeks, your ears, your mouth, all your facial muscles are relaxing. Spend time with each part of your head and allow them to become fully relaxed.

5. Progress down into your neck.

Go through the same process. Relax the right side, now the left side of your neck.

6. Now, progress down into your chest and stomach.

Listen to your heart as it beats rhythmically. As you relax, let your heartbeat slow down, beating calmly, peacefully.

Take a slow, deep breath into your diaphragm. Now let it out slowly, completely. As you do, let all your chest muscles relax, now your stomach muscles.

7. Allow your arms to relax.

Relax your arms, starting with your fingers. Let the relaxed feeling move into your forearms, your elbows, your upper arms. Relax every muscle in your arms, hands, and fingers.

8. Relax all the muscles in your pelvis and groin.

Let your hips relax, your butt relax, all the muscles around your groin relax.

9. Finally allow your feet and legs to relax.

Starting with your toes, then let the relaxed feeling move into your feet, your ankles. Let the relaxed feeling move into your shins and your calves. Allow it to move into your thighs.

You are now totally relaxed and completely comfortable. Feeling just like a limp, rag doll. Completely relaxed, so very comfortable and at peace with yourself and the world around you.

Now it's time to give yourself the positive suggestions and affirmations you created to help you with whatever you are using this self-hypnosis session for.

Chapter 11

Neuro Linguistic Programming

"The quality of your communication shapes the quality of your life. Every cell in your being aligns with what you declare." - **Niurka**

Neuro Linguistic Programming or NLP is another powerful mind technology that you can use to work with your subconscious mind.

Simply by learning the basics of NLP, you will be able to communicate more effectively with yourself as well as others.

You can use your improved communications to motivate you. It can help you think more positively about yourself and the world around you. This new mindset or perspective can enable you to take the actions needed to create the life you desire.

What is NLP?

NLP is a skill people from all walks of life learn to use to model excellence in other people. Through this modeling process, they can put the positive qualities they are emulating to use in their own lives quickly and easily.

Therapists, educators, trainers, and coaches have been using NLP for over 40 years to help people work through issues and improve performance. These are issues like overcoming a neurosis such as the fear of flying or helping people relax and so they can have a successful romantic encounter.

NLP is one of the best-kept secrets of success for athletic improvement of all levels. Athletes have used NLP for issues from improving their free throw ability to working through mental barriers that are holding them back from success.

Professional and Olympic Athletes, Fortune 500 companies, Grammy-winning musicians, Oscar-winning actors, and Special Forces operators, have all used NLP. People also use it in therapeutic settings to help them eliminate bad habits, depression, guilt, stress and phobias.

Defining NLP

The term NLP represents Neuro Linguistic Programming:

Neuro - The nervous system through which all experience is received and processed. Simply speaking, these are the five senses: sight, sound, hearing, taste and feel (tactile).

Linguistic - The language and nonverbal communication systems that every person uses in one manner or another to code and decode (give meaning to) the information in those neural representations.

Programming - Aligning those communication and neurological systems in a way that motivates a person to take the action needed to achieve specific desired goals and results.

How Does NLP Work?

"There is no such thing as reality, only our perception of it." - **Becky Mallery**

As we grow, we develop views and perspectives about our experiences that allow us to make sense of the world around us. After a while, the way we look at the world becomes habitual. We begin to "frame" our new experiences through this habitual way of seeing the world.

Some frames are valuable and help us live life to the fullest. Others are disempowering and hold us back. To better understand this, let's take a deeper look at frames.

Frame of Reference

How you frame an experience gives it significance. This is known in psychology as your frame of reference.

Something that you frame as important will stay with you for a long time. How you frame something is why an event that happened 20 years ago can still affect what you think or believe about yourself today. This is why you can still remember what you think your parents or teachers said to you long into adulthood.

For example, you're 10 years old. You raise your hand to answer a math question. If you answered it correctly, you got your teacher's approval. You framed this experience as a positive one. An event that showed you were smart in math.

On the other hand, if you got the answer wrong, you might have gotten a disapproving look from your teacher. You might have heard snickering from your classmates. You framed this experience as being a negative one. An experience that showed you that you weren't smart in math.

No matter whether the frame is real or not, the frame you give an experience affects how you think about yourself, sometimes for a lifetime.

That's unless you can learn to reframe it.

Reframing

Reframing is being able to see an event or activity from a different point of view.

Let's say you lost your job.

Losing your job could make you feel unhappy. You might feel upset because you had that job for 5 years. You were accustomed to the work there. It was easy. The pay was enough to get by and pay your bills.

But, by using reframing, you can look at losing your job from a different perspective. You can then use your new perspective to create new opportunities.

Let's reframe losing your job.

First, you have the time to look for a better job. A job you'd really like to have. One that challenges you and makes you stretch your abilities.

With the new job you might be able to earn more money. You might like your co-workers better. Your new job may be a lot more in line with what you would consider your dream job.

Now think about your old job for a moment and use reframing to contrast it with a new job.

While you were accustomed to the work in your old job, you weren't learning anything new. A new job might give you an opportunity to gain experience and grow personally and professionally. It might give you the opportunity to travel to various parts of the country or around the world. It might allow you to be part of a bigger team.

Reframing allows you to see what's happening from a new perspective. This new perspective can help you see this event as an opportunity, instead of a problem.

When Neurology and Linguistics Meet

When you have a certain "frame," let's say "I like learning and am good at studying," your brain receives two sets of representations about that frame.

The first is the symbols you make up to represent the frame.

In this example, you might have a certain word, such as "success" related to this frame. Perhaps you see a certain picture that relates to "learning" and "studying." But what's interesting is that along with those symbols, you'll also relate a certain feeling in your body to that frame.

Often, that feeling is accompanied by input from other senses like a sound or a taste. When you take the time to analyze it, you can usually find a location in your body where that feeling is coming from.

So, when the symbol for "I like learning and am good at studying" comes up in your mind, a feeling comes simultaneously somewhere in your body. This is how you know with certainty that "I like learning and am a good at studying." It comes from this programming of symbols (linguistic) and feelings (neuro). You have used them to "program" that feeling into you.

This is a key belief in NLP.

We organize our brain's programs (what we do habitually) by how we systematically communicate to ourselves and to others. Since much of this occurs outside of our awareness, in the subconscious mind, NLP uses techniques that work simultaneously with both your conscious and subconscious minds.

Utilizing Reframing

You can utilize reframing to get to the root of all your problems: decision-making.

Why do I call decision making the root of all problems?

Well, because, for most people, decision-making is where all their problems start.

They want to get healthier BUT instead of going to the gym, they decide to go straight home and watch TV while sitting on the couch.

Why? Because making time for the gym is hard.

They have defined going to the gym as a problem.

They want to have more job satisfaction and think that opening their own business could provide that job satisfaction BUT they decide to stay at a job they don't really like all that much. A job that stifles their creativity and growth potential.

Why?

Because they need the money their present job provides them to maintain their lifestyle. Opening their own business could put a huge dent in their cash flow. Plus, it's risky.

They have defined opening a new business as a problem.

They have thought about getting a master's degree BUT they decide not to pursue it. They understand that it could help them in their current job and earning a new degree would make them feel better, but they decide it's too hard.

Why?

Because they already work up to 60 hours a week and are afraid that going back to school will cost them too much time and money.

They have defined going back to school as a problem.

These three decisions are all focused on a problem.

But what if you could make all these problems go away?

You can.

How?

By putting a different frame on your decisions.

So, how does a frame get rid of your problems?

When any issue comes up, there are always two frames from which you can choose.

You've already been introduced to the first one. It's called the problem frame.

The problem frame focuses on what went wrong or who or what is wrong. Another name for the problem frame is the blame frame. This is blaming people and things outside of you for any issues that come up.

To help you recognize the blame frame we'll use our examples.

Making time for the gym is a problem.

The blame here is not having enough time to dedicate to your health. The problem is something outside of you is stealing your time.

Opening your own business is a problem.

The blame here is a new business could take away from your cashflow and negatively impact your lifestyle.

Earning your master's degree is a problem.

The blame here is that your present job takes too much time and provides too much income to walk away from.

The real question in all these situations is, "Who's to blame for this?" or "What's to blame for this?"

But there is a second frame, known as the outcome frame.

The outcome frame focuses your imagination on your desired result, on what you want, on the possibilities this issue or change brings. Focusing on possibilities attracts the energy and resources you need. This gives you the motivation to take the action you need to do so you can get done what needs to be done.

Going back to our examples.

I want a healthy body.

I imagine myself feeling and looking healthy and going to the gym helps me do this.

I want to open my own business.

I imagine having a thriving business and I look for ways I could open my business.

I want to earn a master's degree.

I imagine what having a master's degree will do for my career and begin researching options for taking courses online.

Here's the difference.

Instead of focusing on the problem, you ask the question:

"What do I want?" or

"What do I have that can help me succeed?" or

"Who can I ask to help me succeed?"

This frame gets your mind searching for a solution. By focusing your mind on the solution, instead of being stuck, you start figuring out what you need to do to achieve your desired result.

For example, let's say you are struggling with speaking to groups.

Your problem statement might be, "Why do I always struggle so much when I speak in front of a group?"

This, of course, is a negative outcome statement. It focuses your attention on the problem.

To turn this around, ask yourself, "How would I like to feel when I present to groups?"

Your response might be, "I want to be relaxed and confident so that I can share my information with my audience."

Now you're focused on what you need to do to get the outcome you are looking for. You've successfully reframed your problem into an achievable outcome!

This is the power of frames!

Chapter 12

Mindfulness

"The best way to capture moments is to pay attention. This is how we cultivate mindfulness. Mindfulness means being awake. It means knowing what you are doing." - Jon Kabat-Zinn

---

The roots of mindfulness can be traced back thousands of years to the ancient teachings and practices of Buddhism. Buddhist monks, seeking spiritual enlightenment and liberation from suffering, developed mindfulness meditation as a principal component of their spiritual path.

The concept of mindfulness, known as "sati" in the Pali language of early Buddhist texts, refers to the practice of intentionally and non-judgmentally paying attention to the present moment. It involves cultivating a heightened awareness of one's thoughts, emotions, bodily sensations, and the surrounding environment.

Mindfulness meditation serves to develop this state of mindful awareness.

What is Mindfulness?

Let me ask you a question. What do you think of when someone mentions meditation?

Do images of religious monks sitting together in circles and chanting "ohhmm" form in your mind?

Maybe you're a little skeptical of meditation because you feel that it would take you hours each day trying to get the technique right. Well, think again. But, before we look at what mindfulness is, let's look at what mindfulness is NOT.

This is NOT Mindfulness

Mindfulness is not:

- Spending time thinking about the world and all the wonders you can find in it

- Contemplating the great, deep mysteries of life

- Some magical practice that brings good luck into your life

- A spiritual practice reserved only for those in highest connection to their Creator and can dedicate hours to every day

- Daydreaming or fantasizing about what your perfect life looks like

- Mindfulness is NOT Mystical.

Yes, meditation and mindfulness in the modern world may have some of these mystical connotations, but mindfulness is much simpler.

In simple terms, mindfulness and meditation are ways to achieve a calmer state of mind. Mindfulness is essentially a sub form of meditation.

Although mindfulness originated with non-western practitioners ages ago, today it's practiced and studied by scientists, psychologists, and doctors all over the world.

Defining Mindfulness

Now that we have a better understanding of what mindfulness is not, let's consider what the practice of mindfulness actually is.

Jon Kabat-Zinn is credited with bringing the popularity of mindfulness to the United States. He continues to develop his understanding of mindfulness through his Mindfulness-Based Stress Reduction program at the University of Massachusetts Medical Center.

Jon defines mindfulness like this:

"Mindfulness means paying attention in a particular way; on purpose, in the present moment, and non-judgmentally."

Two different Chinese calligraphy characters make up the term "mindfulness." The top character means "presence" and the bottom one means "heart." Therefore, mindfulness literally means "presence of heart."

Mindfulness is simply experiencing and focusing on the present moment.

Here's why learning to focus on the moment is so important. Whether conscious of it or not, most people always focus on things and events of the past or future.

For example, most people constantly look forward to the last day of the workweek while at work, and what their weekend plans will bring. They look forward to their summer vacations. They look forward to the holidays. They look forward to their birthdays. They look forward to the end of the day when they get to crawl back into bed again.

As you can see, we, like most people, never focus entirely on the present moment. We let our minds float into the future and ruminate about the past.

This is how to best utilize mindfulness. Mindfulness is a way to learn how to focus on the present moment, to create an awareness, an acute, keen and specific consciousness of ourselves and to center ourselves right now in the present, not sometime in the future or past.

Step-by-Step Guide to Mindfulness Meditation

Practicing mindfulness meditation does take time, effort, and patience. That being said, the advantages you receive from this practice certainly outweigh the costs.

Here's what you need to remember. You don't need to spend hours a day practicing mindfulness. You can achieve excellent results by practicing as little as 20 to 30 minutes per day.

Follow these simple steps to practice mindfulness meditation:

1. Find a quiet place.

With mindfulness, or any form of meditation, you can practice anywhere at any time. However, the easiest place, especially when you're first starting out, is a quiet one, free from any distractions.

Places that you may be able to practice include in your car, at home, in a quiet spot in the park, or even at work if you have an office where you can close your door.

2. Sit in a comfortable position.

Many people believe you must sit in a lotus-style type of position. This is where you sit cross-legged with your feet resting on opposing thighs. Some people believe you must sit cross-legged, Indian style.

There is no correct position.

If you are comfortable doing it, you can choose to sit cross-legged with your feet resting underneath your knees. You can also sit in a desk chair, on the floor with your legs bent or out in front of you, or any seat where you can remain comfortable for a few minutes.

3. Focus on your breath.

Close your eyes and take several slow breaths. Focus on inhaling slowly and deeply into your diaphragm and exhaling completely. Do this several times.

As your muscles and cells receive the oxygen from your deep breathing, you'll become more relaxed. This will allow your mind to calm and slow down.

As you sit, continue to pay attention to the air as it comes in and goes out.

4. Notice physical sensations.

As you notice your breathing, also pay attention to the physical sensations in your body.

Notice the weight of your body in your seat. How your bottom feels as you sit there. Make a note if you're leaning further to the right or to the left. Pay attention to the temperature in the room and your body's temperature.

How do you feel? Do you feel warm, cool, or somewhere in the middle?

5. Notice the noises around you.

Don't pay attention to any particular noise. Just let the noises be and notice their presence.

No matter whether it's the whistling of the fan, the hum of the air conditioner or heater, or the chatter of people's voices outside, the birds outside, just notice the sounds around you.

6. Notice when your mind wanders.

During any type of meditation, it's perfectly normal, especially for beginners, for your mind to wander off in thought. You may start thinking about this, or that. There are several things that can pop into your mind.

When your mind wanders, as it always does, just notice it. Now quietly bring your attention and awareness back to your breathing. Pay attention to the air slowly going in and going out of your lungs.

7. End your mindfulness session when you're ready.

At the end of your mindfulness session, slowly open your eyes. Stretch.

Now think about your experience. What do you feel at this moment in time? Do you feel more relaxed, calmer, more focused?

When you first start mindfulness meditation, within a few moments you'll find your mind wandering. It's difficult, if not impossible, to keep thinking about your breathing.

Don't be discouraged. This is normal.

As you practice, it will become easier for you to clear your mind and focus on your breath. Like learning all skills, the key is to practice consistently.

Take the Mindful 20/30 Challenge

We face thousands of decisions each day. These decisions challenge us from the moment our alarm clock startles us awake in the morning until the time our minds drift off to sleep at night.

Studies have found that the average adult makes about 35,000 decisions in a single 24-hour period. These are decisions like:

Do I hit snooze on the alarm clock or get up and do the exercise program I promised myself I would?

What should I eat for breakfast, oatmeal or bacon and eggs?

What clothes should I wear today? Should I wear the green or the blue shirt? Which pair of shoes match that outfit?

Should I take the dog out for a short walk and get some exercise before leaving for work?

Should I go into work or call in sick?

That guy just cut me off, should I yell and scream at him?

Decisions constantly threaten to overwhelm us. We face problems, tasks, drama, stress, workloads, good news, unwelcome news, indifferent news, world news, and mass amounts of information.

This data piles up daily. As a result, we routinely experience an information overload that affects our minds, bodies, and souls!

Living in the present, being mindful, and deliberately practicing mindfulness is one way in which we can tell our bodies and our whole being to take a moment and slow down.

But there is a constant battle going on inside of us. One that understands the need to take time for ourselves and the other telling us we're literally so busy we don't have time to slow down.

What can you do?

You must make the decision to stay true to your real self. The person who lives here and now, in the present.

Practicing mindfulness daily is a way in which we can literally free ourselves of distractions. However, making the decision to practice mindfulness is only the first step.

After you make the commitment to the daily practice of mindfulness, you need to develop it into a habit. This takes effort.

With a strong commitment and a solid resolve to make it happen, you're halfway there. You've almost won the battle to creating a calmer, healthier, happier self!

The last step is to take action.

You can use this challenge to do just that.

Taking the Challenge

If you feel up to it, challenge yourself to put mindfulness into practice for 20 minutes each day for 30 consecutive days.

Here's what you need to do:

1. Put an appointment in your calendar for at least 20 minutes every day.

2. Locate a quiet space. You want a spot where you can meditate peacefully while experiencing the power of being mindful.

3. You can practice mindfulness at the same or different times on different days. You don't have to practice mindfulness at the same time each day for these 30 days. You DO have to take the time for mindfulness each day.

4. After practicing, write about your experience in your journal.

5. Every week, and after the 30 days is over, read what you've written in your journal. Reflect on how mindfulness

has impacted, affected, or changed you and write this into your journal.

Did you complete all 30 days?

If not, it's okay. Simply try to finish it again!

Some Final Words About Mindfulness

"Now is the future that you promised yourself last year, last month, last week. Now is the only moment you'll ever really have. Mindfulness is about waking up to this." - **Mark Williams**

Researchers and scientists who have studied the effects of practicing mindfulness often deliberately choose people who haven't had any exposure to mindfulness or meditation training.

Why?

They want to find out if a person with no prior training can reap the benefits of beginning mindfulness.

What did they find?

If you practice, anyone can benefit from practicing mindfulness.

Proven benefits include:

- Lowering your stress hormones

- Increasing your energy levels

- Better mental focus and stamina
- Enhanced memory and attention span
- Boost to the immune system
- Lowering blood pressure and heart rate to normal ranges
- Decreases fatigue
- The brain functions better
- Lower reaction to emotions and pain
- Heightened self-awareness
- Increase in compassion and empathy towards others
- Better sleep quality
- Lower rates of depression

While it can be difficult to understand how sitting in a quiet space for a few minutes each day can impact your life, just a few days of consistent practice can completely change your perspective. After practicing for merely 30 days or so, you will begin to notice how mindfulness benefits your physical, emotional, and behavioral health.

Mindfulness is not just for "new-agey" or deeply spiritual people. Anyone can use it.

It just takes a small level of commitment and consistent practice. With practice, over time, you will notice greater mental control and feel healthier.

The great news is you don't have to take this on faith. There are tons of research to help you with your practice.

When you combine this research with a deliberate, purposeful action of making mindfulness part of your daily routine, you, like millions of others, can use mindfulness to transform the quality of your life.

You can practice mindful breathing, mindful listening, mindful observation or choose one of the many other forms of meditation. Channeling your focus and attention for just 10 to 30 minutes each day is enough to set you on the path to a happier, healthier life.

If you'd like to have your own mindfulness experience, you can take my 5-Day Mindfulness Challenge by going to MindfulMindHacking.com

Discover Your Path to Personal Transformation - Free Course!

JOIN US FOR A FREE Course: Discover Your Path to Personal Transformation!

Are you ready to unlock your true potential? We invite you to enroll in our **FREE course** designed to open the door to personal growth through mindfulness, Neuro-Linguistic Programming (NLP), and visualization.

In this course, you'll learn how these powerful tools can help you manage emotions, enhance communication skills, and set effective goals. With practical exercises and techniques at your fingertips, you'll be empowered to take charge of your personal transformation.

Start your journey of self-discovery and create a life filled with meaning and fulfillment!

Don't miss out on this incredible opportunity—follow this link – https://blackbeltbreakthroughs.com/navigating-the-path-of-living-an-intentional-life/

or scan the QR code to begin your transformative journey today!

Let's travel this adventure together!

Chapter 13

How to Know If Your Reprogramming Efforts Are Working

"You don't make progress by standing on the sidelines, whimpering and complaining. You make progress by implementing ideas." - **Shirley Chisholm**

Reprogramming the subconscious mind can be a formidable task.

As you've seen, the contents of your subconscious often remain hidden from your conscious awareness. Unlike your conscious thoughts, which you can easily observe and scrutinize, the subconscious operates beneath your conscious radar. It wields influence over your thoughts, emotions, and actions in profound ways.

Reprogramming the subconscious requires a systematic and strategic approach. It involves unearthing and confronting the beliefs and patterns that no longer serve you, then replacing them with new, empowering ones.

To do this, you need to develop a keen sense of self-awareness. You need to be able to catch any self-sabotaging behavior before it starts to take over and gets out of hand.

While it may seem daunting, with the right tools and techniques, like the ones in this book, anyone can build this awareness. They can then use it to cultivate a new, positive subconscious programming that enhances their life experience.

Signs That Your Reprogramming is Working

There are some clear signs of progress that you may recognize:

You begin feeling stronger, more confident, and happier.

You find yourself more willing to take risks and face challenges.

Your dreams and goals don't seem overwhelming anymore – just exciting.

You feel a deeper sense of inner peace, as if inner conflicts are dissolving.

You attract more opportunities to expand and grow in every area of your life.

In short, you'll know when changes are taking place in your subconscious mind because you'll notice a shift in both your inner and outer being.

The evidence is usually undeniable.

You just need to look for it.

Chapter 14

Consistent, Persistent Reinforcement

"Energy and persistence conquer all things." - **Benjamin Franklin**

Ben Franklin's quote encapsulates the essence of the subconscious reprogramming process. It serves as a reminder that achieving lasting change requires steadfast dedication and unwavering perseverance.

When embarking on this journey it is vital to approach it with patience and a steadfast long-term perspective. This is the only path to a successful transformation.

It is crucial to remember that this process cannot be done quickly. It is not magic. Like physical exercise builds your body, you need to build your mind's strength through regular mental and emotional workouts.

Consistency plays a pivotal role in this process. Regularly practicing the methods in this book, whether it be it affirmations, visualization, meditation, or other techniques, fortifies the new messages and embeds them deeply in your subconscious mind.

Consistency enables your subconscious to gradually embrace and integrate these positive messages, gradually replacing your old patterns and beliefs.

Keep in mind that it is natural to encounter feelings of doubt or impatience during this journey. Nevertheless, it is crucial to maintain a steadfast commitment and persistence, even when your results may not be immediately evident.

Like drops of water, slowly drilling a hole in a rock, reprogramming the subconscious mind is a gradual process that unfolds over time. It takes time to instill more positive messages into your subconscious mind.

But with consistent practice, you set the foundation for the changes you desire. You can change your life.

As the process of reprogramming your mind begins to take effect, you'll often experience a surge in motivation and encouragement. Positive shifts in thoughts, emotions, and behaviors become increasingly evident, bolstering your confidence in the approach's efficacy.

These tangible outcomes serve as a driving force to propel you on your journey towards self-improvement and positive changes.

These transformations can be profound and long-lasting. They will engender a fundamental shift in your perspectives, attitudes, and decision-making processes.

Although these changes may take time to fully manifest, they are certainly worth the investment you make in time and effort. They will empower you by fostering inner harmony. This will allow you to align your actions with your deepest desires.

As soon as these transformations become apparent, you'll feel motivated to keep moving forward, but until that happens, stick with it.

Believe that these changes are lifelong, powerful, and well worth waiting for!

About the Author

Are you feeling a little lost and unsure of your path in life?

We've got just the solution for you!

Meet Wil Dieck, the mastermind behind Mindful Mind Hacking.

With his extensive background in hypnotherapy, NLP, martial arts, and mindfulness, he's got the tools you need to unlock your true potential and create the life you've always dreamed of.

And who doesn't want that?

Wil's forty years of experience have helped countless clients and students achieve their goals and break free from any barriers holding them back.

So, whether you're looking for online courses, books, one-on-one coaching, or group presentations, Wil's got you covered.

Don't wait any longer - visit mindfulmindhacking.com and start hacking into your success today!

Other Books by Wil Dieck

Modern Mindfulness: A Beginners Guide on How to Find Peace and Happiness in a Busy World

The Secrets of the Black Belt Mindset: Turning Simple Habits Into Extraordinary Success

Mastering the Mind, Body and Spirit: Secrets of Black Belt Peak Performance

Mindful Mastery: Find Focus, Get Unstuck, and Drop Into the Peak Performance Zone

NLP - UNLOCK YOUR DREAMS: A Beginners Guide to Neuro Linguistic Programming (3rd edition)

Made in the USA
Columbia, SC
04 January 2025

09311f9c-2658-4ed3-8c6a-7bcb839ba2dbR01